DATE DUE			

9793

523.1 Moore, Patrick.
MOO
The universe for the beginner.

**MESA VERDE MIDDLE SCHOOL
POWAY UNIFIED SCHOOL DISTRICT**

378867 01208 519984 02246E   14

THE UNIVERSE
FOR THE BEGINNER

THE UNIVERSE
FOR THE BEGINNER

Patrick Moore

The title page picture is an artist's impression
of the planet of a red giant. The red star shines
down on the planet's barren landscape. A
satellite of the planet is seen to the upper right.

Published in North America by the Press Syndicate of the University of Cambridge
40 West 20th Street, New York, NY 10011-4211, USA
Published throughout the rest of the world by George Philip Ltd.
59 Grosvenor Street, London W1X 9DA

© Myfanwy Hester Woodward 1990

First published 1990
First Cambridge University Press edition 1992

Printed in Hong Kong

A catalog record for this book is available from the British Library

ISBN 0-521-41834-8 hardback

CONTENTS

CHAPTER	1	OUR HOME IN SPACE	6
CHAPTER	2	THE SUN'S FAMILY	8
CHAPTER	3	OTHER SUNS	10
CHAPTER	4	THE MESSAGE OF LIGHT	12
CHAPTER	5	HOW LIGHT BEHAVES	14
CHAPTER	6	HOW THE STARS SHINE	16
CHAPTER	7	GIANTS AND DWARFS OF THE SKY	18
CHAPTER	8	EXPLOSIONS IN THE SKY	20
CHAPTER	9	BLACK HOLES	22
CHAPTER	10	THE SIZE OF THE GALAXY	24
CHAPTER	11	OTHER GALAXIES	26
CHAPTER	12	DIFFERENT KINDS OF GALAXIES	28
CHAPTER	13	THE EXPANDING UNIVERSE	30
CHAPTER	14	THE STORY OF QUASARS	32
CHAPTER	15	THE EDGE OF THE UNIVERSE	34
CHAPTER	16	THE BIG BANG	36
CHAPTER	17	THE IN-AND-OUT UNIVERSE	38
CHAPTER	18	LIFE IN THE UNIVERSE	40
CHAPTER	19	TRAVEL THROUGH THE UNIVERSE	42
CHAPTER	20	THE END OF THE UNIVERSE	44
		TEST YOUR SKILL!	46
		ANSWERS	48

CHAPTER 1
OUR HOME IN SPACE

Look into the sky during daytime, and you will see the Sun. At night you will see the stars, and for part of each month you will also see the Moon. But have you ever asked yourself: 'How big is space?' And have you ever asked yourself how the Sun and the stars were born?

The universe—that is to say, everything we know—is so large that we cannot hope to understand its size. Even the Moon, which is much the nearest body in the sky, is so far away that you would have to fly ten times round the world before you had covered a distance the same as that between the Earth and the Moon. In astronomy, we have to become used to very great distances and very long periods of time.

We have learned a great deal during the past few years, partly because we have built better telescopes and partly because we have found out how to travel into space. Men first reached the Moon in 1969, and rockets have been sent out to the planets. We can now 'see' further into space than we have ever been able to before.

In this book I am going to try to give some idea of what we believe the universe to be like. First, we must deal with our own Earth and its neighbours in space.

The Earth is a *planet*—a world nearly 8000 miles (12,800 kilometres) across, moving round the Sun. It takes $365\frac{1}{4}$ days to make one full journey, and this of course is the 'year'. The Earth itself spins round once in 24 hours, which is why we have day and night. The Sun can shine upon only half the Earth at any one time, so that when we are on the side of the Earth which is opposite to the Sun we are in darkness. Because the Earth spins from west to east, the whole of the sky seems to move from east to west.

The Sun is a star, and all the stars you can see on any clear night are themselves suns. Many of them are much larger, hotter and more powerful than our Sun, but they are so far away that they look like nothing more than tiny points of light. The distance between the Earth and the Sun is 93 million miles (150 million kilometres). If we give a scale model, and put the Earth and the Sun at opposite ends of this line ——————, the nearest star will have to be put about 4 miles ($6\frac{1}{2}$ kilometres) away.

This picture of the Earth was taken from Apollo 11 on its way to the Moon. The outlines of the continents can be seen.

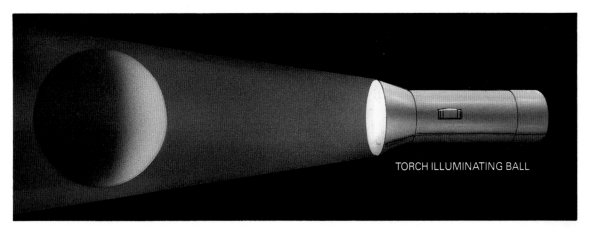

TORCH ILLUMINATING BALL

Just as a flashlight lights up one half of a ball, so the Sun lights up one half of the Earth.

The Earth is not the only planet moving round the Sun. There are eight others, some of them larger than the Earth and some of them smaller. They have no light of their own, and shine only because they are lit up by the rays of the Sun—just as a white ball will if you shine a torch upon it in a dark room. Some of the planets look brilliant, and most of them have *satellites* moving round them. We have one natural satellite—the Moon, which is not as important as we often think. Like the planets, it shines by reflecting the light of the Sun.

The Sun is one of about 100,000 million stars making up the star-system which we call the *Galaxy*. Unlike the planets, the stars shine by their own light. They cannot be seen in the daytime because the sky is too bright; at night, you may be able to see over 2000 of them with the naked eye (that is to say, without using a telescope or a pair of binoculars). The rest are too faint, because they are so far away.

At even greater distances we can make out dim, misty patches which we know to be other star-systems or galaxies. Each of these galaxies is made up of stars. The brightest galaxy we can see from America is called the Great Spiral in Andromeda. The picture of it shown here was taken with one of the world's largest telescopes; each point of light in the photograph is a sun.

The Great Spiral in Andromeda, photographed with the 200-inch reflector at Palomar, California.

You can see, then, that the Earth is of no real importance in the universe, and is important to us only because we live on it; it is our home in space. Thousands of years ago, people believed the Earth to be flat, and to lie in the middle of the universe. We know better today.

CHAPTER 2
THE SUN'S FAMILY

Most of this book will deal with the stars and star-systems, but I must begin by saying a little about the Sun's family, which we call the Solar System (from the Latin word *sol*, meaning 'sun').

The nearest of all bodies is the Moon, which is much smaller than the Earth. If the Earth is taken to be the size of a tennis ball, the Moon will be no larger than a table-tennis ball. It has no air, so we could not live there; the men who have been to the Moon have had to stay inside their spaceships or their space-suits. The first trip was made by two Americans, Neil Armstrong and Edwin ('Buzz') Aldrin, in their spacecraft *Apollo 11*. As they walked about on the Moon, they could see the Earth shining down at them from the black sky. The last lunar journey was made in 1972, but we expect people to go back there within the next few years.

There is not much air at a height of more than about 500 miles (800 kilometres) above the ground, and this means that ordinary aircraft cannot be used for space travel. We have to use rockets. I will not say much about them here, because I have talked about them in the second book of this series, *Space Travel for the Beginner* (1992). We have also sent up man-made moons, or *artificial satellites*, some of which carry telescopes. As they are moving above the top of the air, these telescopes can 'see' very clearly—they do not have to worry about clouds and fog.

The Moon moves round the Earth, making one complete journey in 27.3 days; it is always close to us, and stays together with us as we travel round the Sun. The other planets move round the Sun, and are much further away, so that a rocket takes much longer to reach them.

The Solar System is divided into two parts. First we have four small planets: Mercury, Venus, the Earth and Mars. Mercury is never very easy to see because it is so close to the Sun (only 36 million miles, or 58 million kilometres, away from it), and it has no air, so it can have no life. Venus is almost as large as the Earth, and is very brilliant indeed. However, it is not like the Earth. It has a dense air which we could not breathe, and is very hot, with

(*Left*) The Earth and the Moon, shown to the correct scale. The Moon is very much smaller.

(*Below left*) 'Buzz' Aldrin stands on the Sea of Tranquillity during the *Apollo 11* mission.

(*Right*) The large diagram shows the path of Mars, and the outer planets to the same scale. The other shows the inner planets' paths round the Sun.

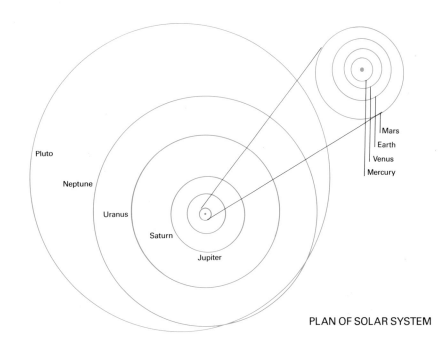

PLAN OF SOLAR SYSTEM

a surface temperature of 480 degrees Celsius. It is not likely that anyone will visit it, but unmanned rockets have landed there and have sent back pictures.

Beyond the Earth we come to Mars, the red planet. It is smaller than the Earth, though larger than the Moon; its air is thin, and it is very cold. Its poles are covered with ice, but there are no seas, and the rockets which have landed there have found no signs of life. However, Mars is one of the closest of the planets, and when nearest to us is only about 150 times as far away as the Moon. It is quite likely that humans will go there within the next 40 or 50 years.

Outside the path of Mars there is a wide gap, in which move thousands of very small worlds called *asteroids*. Next come four giant planets: Jupiter, Saturn, Uranus and Neptune. They are not solid and rocky, but have surfaces made up of gas. Jupiter, the largest of them, is so big that it could hold over a thousand bodies such as the Earth.

Telescopes give good views of these four planets—particularly Saturn, with its lovely set of rings. The two outer giants, Uranus and Neptune, were not known in ancient times; you can just see Uranus with the naked eye, but Neptune is too faint. All four have been bypassed by unmanned rockets which have sent back splendid pictures of them. There is one more distant planet, Pluto, which is smaller than the Moon.

We also find comets, which are made up of ice, gas and 'dust'; sometimes you will see shooting-stars, which are not really stars but simply tiny specks of material burning away in the Earth's upper air. I have described all these in *Astronomy for the Beginner* (1992), the first book in this series.

Our rockets can travel to the edge of the Solar System, but to send a rocket to another star would take thousands of years. So when we study the greater universe, we have to use instruments either on the Earth, or carried in man-made satellites moving round us.

CHAPTER 3
OTHER SUNS

Though the stars are suns, no telescope will show them as anything but specks of light. It is only the Moon and planets which show disks upon which we can see markings. Of course, we can see the Sun clearly—but *never look straight at it*, because you will blind yourself; and above all, never look at it through any telescope or binoculars. This is very dangerous, so I would ask you to take great care.

The stars make up patterns or *constellations*. All the stars are moving around in space, but they are so far away that their individual movements are too slow to be noticed; the constellations we see today are the same as those which must have been seen by Julius Caesar, or even by people of the Stone Age. It is only the members of the Solar System which wander slowly about from one constellation to another. Of course, the whole sky seems to turn round from east to west once in 24 hours, but this is only because the Earth itself is spinning round.

We measure the distances of the Sun, Moon and planets in miles or kilometres, but the stars are so far away that we have to use a different unit. (We could use miles, of course, but it would be awkward, just as it would be awkward to give the distance between London and New York in inches or millimetres. Work it out, and you will see what I mean!) The astronomer's unit is known as the *light-year*.

Light moves at a speed of 186,000 miles (300,000 kilometres) per second, so it can flash from the Earth to the Moon in $1\frac{1}{4}$ seconds. The Sun's light takes $8\frac{1}{2}$ minutes to reach us, but the light from the nearest star takes over four years. In this book I will be giving most distances in light-years. If you want to change them to miles, multiply by 6 million million; for changing to kilometres, multiply by $9\frac{1}{2}$ million million. The most distant objects we have found at present are over 13,000 million light-years away, which gives you some idea how large the universe really is.

The stars you can see with the naked eye belong to our own Galaxy. It is a flattened system, as shown in the

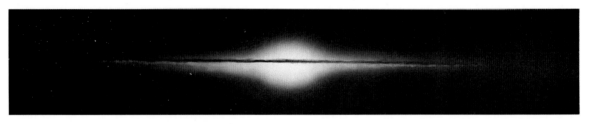

The shape of our Galaxy, seen edge-on. The Sun lies well away from the centre.

picture; if you take two fried eggs and put them together back to back, you will have some idea of the Galaxy's shape. The Sun lies well away from the centre of the Galaxy (in the 'white' of our model egg!). If you look along the main thickness of the system—that is to say, either straight towards the yolk of the egg in our model, or else straight away from it—you will see many stars in almost the same direction. This makes up the lovely shining band of the Milky Way, which is beautiful when seen against a dark sky, though it is not easy to see from a town or city as it is drowned by artificial lights.

Astronomy becomes much more interesting when you learn some of the most important constellations. Once you have found them, you will always be able to recognize them again, because they never change. For example, there is the Great Bear, which never sets over America, so it is always on view whenever the sky is dark and clear. The seven stars making up the main pattern are fairly bright, and two of them (the 'Pointers') show the way to the Pole Star, which lies close to the north pole of the sky and does not seem to move as the Earth spins round. Of course, it is not really standing still; it just happens to lie in line with the Earth's axis of rotation. It is 6000 times as powerful as the Sun, and is so far away that its light takes 680 years to reach us. Look at it tonight, and you will see it as it used to be 680 years ago.

Another splendid constellation is Orion, which is not visible all through the year; it is at its best during evenings in winter. There are two brilliant stars, Betelgeux (sometimes pronounced 'Beetle-juice'!) and Rigel; Betelgeux is orange-red, while Rigel is white. There are also three bright stars marking the Belt. Downwards, they point to Sirius, the brightest star in the whole of the sky.

Below the Belt, in Orion's Sword, you will see a misty patch. This is called a *nebula* (from the Latin for 'cloud'), and we know that it is a place where new stars are being born. More than 4500 million years ago, our Sun was born inside a nebula in this way.

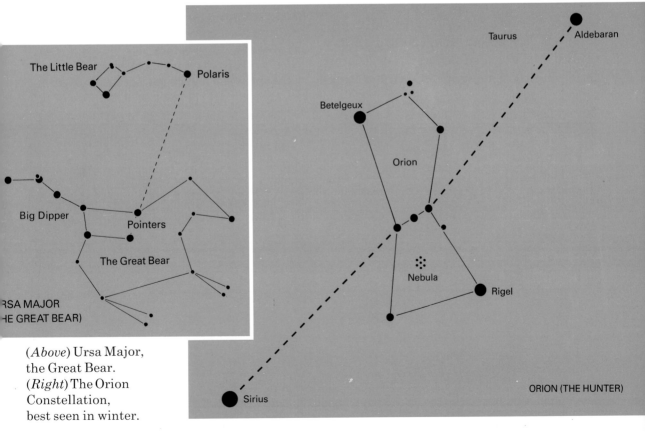

(*Above*) Ursa Major, the Great Bear.
(*Right*) The Orion Constellation, best seen in winter.

11

CHAPTER 4
THE MESSAGE OF LIGHT

An astronomer is always thought of as being a person who uses a telescope. This is true enough, and without telescopes we would know very little about the universe.

A telescope of the kind known as a *refractor* collects its light by using a glass lens or object-glass. The light-rays are bunched up and brought to focus, where an image is formed. We then make the image larger by using a second lens, called an eyepiece, which is really nothing more than a special sort of magnifying glass. With a *reflector*, the light is collected by a curved mirror, and is brought to focus after the rays have been reflected from a second mirror and then sent into the side of the telescope-tube, where the eyepiece is placed. (This is only one type of reflector. There are many others, but we need not say more about them here.) The world's largest telescope has a mirror 236 inches (600 centimetres) across—a 236-inch 'eye'.

Rather surprisingly, modern astronomers do not often look through telescopes. For many years they carried out their work photographically, using the telescopes as giant cameras. Nowadays, photography is being replaced by electronic instruments which are much more powerful. Instead of looking through the eye-end of a telescope, the astronomer is more likely to sit in a comfortable control room looking at the image appearing on a television screen.

Telescopes are used together with instruments called *spectroscopes*, which split up the light-rays and tell us a great deal about the bodies sending us the light. Of course, the larger the telescope, the more light you can collect, and the more you can find out.

(*Below*) A refractor's main lens collects light, the rays are focused, and the image is then magnified by an eyepiece.
(*Bottom*) The reflector's main mirror collects light and sends it back to a secondary mirror.

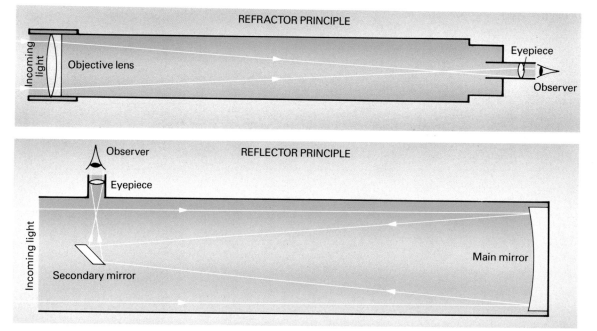

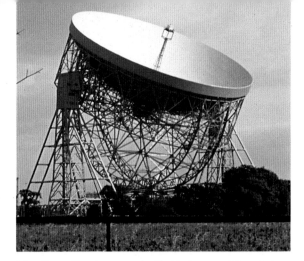

The Lovell telescope in Britain, used to pick up radio waves from the sky.

Light is a wave motion. To show what is meant, throw a stone into a calm pond. Ripples or waves will be set up, and the distance between one wavecrest and the next is known as the *wavelength*. With water, the wavelengths can be measured in inches or centimetres, but the wavelength of light is very much smaller—so small, indeed, that it cannot be measured in the same way.

The colour of light depends upon its wavelength. Red light has the longest wavelength; then come orange, yellow, green, blue and violet (that is to say, the colours of the rainbow). If the wavelength is longer than that of red light, or shorter than violet, the radiation cannot be seen. Visible light makes up only a part of the whole range of possible wavelengths.

Beyond red, we come to what is called *infra-red*. It is easy to detect, even though you cannot see it. Switch on an electric fire; you will feel the infra-red, in the form of heat, well before the bars become hot enough to glow. Rays of still longer wavelength are called *radio waves*, and have to be collected by special instruments called radio telescopes—which are not in the least like ordinary telescopes; they are more like large aerials, some of them 'dishes'. A radio telescope does not give an actual picture, and you cannot look through it, but it can tell us more than we could hope to find out in any other way. The world's most famous radio telescope is at Jodrell Bank, in Britain it has a 'dish' 250 feet (75 metres) across. It is named in honour of Sir Bernard Lovell, the great astronomer who first suggested it could be built.

Many bodies in the sky send out radio waves, but this does not mean that the waves are artificial. As yet, we have no proof of life anywhere in the universe except on the Earth.

If the wavelength is shorter than that of violet light, we have '*ultra-violet*', then '*X-rays*', and finally the very short radiations called '*gamma rays*'. The whole range of wavelengths is shown in the diagram. In astronomy, we have to study them all. If we had to use only visible light, we would be badly handicapped, just as anyone playing a piano would be unable to produce a tune if he could use only the seven notes of the middle octave.

The whole range of wavelengths, from the long radio waves to the very short gamma rays.

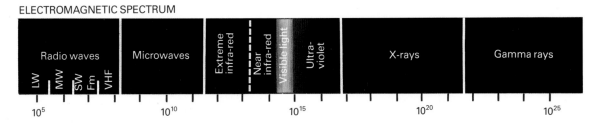

CHAPTER 5
HOW LIGHT BEHAVES

If you listen to an ambulance or a police car which is passing by, sounding its siren, you will be able to notice that the note of the siren changes. When the ambulance is coming towards you, the note is high-pitched. As soon as it has passed by, and has started to move away, the note is increased. This is known as the 'Doppler Effect', in honour of the Austrian scientist who discovered it in 1842.

Light-waves show the same effect. With a body which is approaching, the wavelength of its light is shortened, so that the light appears a little 'too blue'. When the light-source is moving away, the wavelength is lengthened, and the light appears a little 'too red'. The change in colour is too small to be

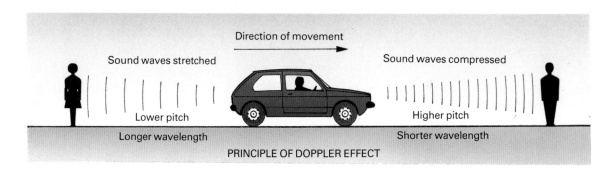

of the siren drops and becomes deeper. The reason for this is that the wavelength changes. Sound, like light, is a wave motion; when the ambulance is coming towards you, more waves per second will reach your ear than would be the case if the ambulance were standing still, so that the wavelength appears shorter than it would otherwise be. When the ambulance is moving away, fewer sound-waves per second reach you, and the wavelength

(*Above*) The Doppler Effect. The siren of a vehicle approaching you sounds high pitched, but becomes low-pitched as it moves away.

(*Far right*) When the spectral lines of a star are shifted towards red, the star is moving away from us. When they are shifted towards blue, the star is moving towards us.

(*Below*) When light passes through a glass prism it forms a spectrum, with violet at the short-wave end and red at the long-wave end.

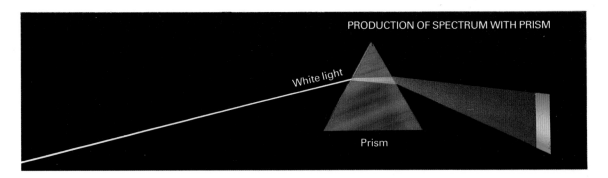

noticed in everyday life (do not expect a yellow traffic-light to turn blue when you drive up to it!), but the effect shows up when we use a spectroscope.

As we have seen, a spectroscope splits up light. What we usually call 'white' light is not really white at all, but is made up of a mixture of all the colors of the rainbow. The first person to realize this was Isaac Newton, in the year 1666. He passed a beam of sunlight through a piece of glass called a prism and found that the different colours in the light were bent or *refracted* by different amounts—violet and blue being bent the most, red being bent the least. The result was that the Sun's light was spread out into a rainbow band or spectrum.

Any body which is sending out light will produce a spectrum. If the source is a hot solid, liquid or dense gas it will give a rainbow spectrum, from red through to violet. But next, let us look at the spectrum of gas which is of much lower density. There will be no rainbow band at all, and all that will be seen are bright lines, separated from each other with no coloured background. We know that each line is the trademark of some particular element or group of elements. Suppose we throw some salt into a flame? Salt contains what we call 'sodium', and this produces two bright yellow lines (among many others). See these lines, and you may be sure that they are due to sodium.

The Sun's surface is dense gas, so that in a spectroscope it gives a rainbow. Above the surface are thinner gases, which give separate lines. Because they are seen against the rainbow background, these lines show up as dark rather than bright, but the positions are the same. In the spectrum of the Sun we can see two lines in the yellow part of the rainbow. These must be due to sodium, so we can tell that there is sodium in the Sun.

Now let us come back to the Doppler Effect. We look at the spectrum of a star, and see two lines in the yellow part of the rainbow. We know exactly where they ought to be. If we measure their positions and see that they have been moved slightly towards the long-wave or red end of the rainbow, it will mean that the star is moving away from us—and the greater the change in position, the quicker the star is moving. This is what is called the *red shift*. If the star is moving towards us, the lines will be moved over towards the short-wave end of the rainbow, and we will have a *blue shift*.

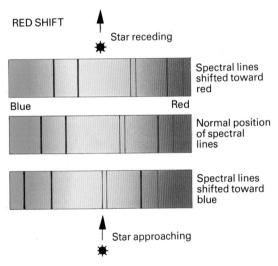

This means that by studying the spectra of the stars, we can tell how they are moving. The same is true of the star-systems or galaxies. Using spectroscopes together with our most powerful telescopes, we can find out the movements of systems which are thousands of millions of light-years away from us.

CHAPTER 6
HOW THE STARS SHINE

In studying the life-stories of the stars, we must begin with the Sun, because it is close to us on the astronomer's scale. After all, it is a very ordinary kind of star, and there are many others in the Galaxy which are exactly like it.

Because the Sun is so hot, with a surface temperature of almost 6000 degrees Celsius, it is natural to think that it must be burning, but this is not true. The Sun shines in a very different way. Deep inside it, one gas is being changed into another, with the result that energy is sent out.

Everything we know—you, me, this book, the Earth, the Sun and the stars—is made up of *atoms*. Atoms are so small that we cannot imagine how tiny they really are, but we have found out a great deal about them, and we know that there are only 92 different kinds of atoms to be found in nature.

These 92 atom-types make up the *elements*. For example, iron is an element; so are gold, silver, lead, tin and sodium. Some elements are best known as gases; our air is made up mainly of two elements, oxygen (which we need to breathe) and nitrogen. Another element is hydrogen, the very light gas which was once used for filling the gas-bags of airships.

Atoms can join together to make up atom-groups or *molecules*. For example, two atoms of hydrogen together with one atom of oxygen make up one molecule of water—remember the famous chemical formula H_2O. It may seem strange to think that two gases can combine to make liquid water, but this is what happens. (Of course, H_2O can also exist as ice or steam, depending upon its temperature.)

Hydrogen is the commonest of all the elements as well as being the lightest. In fact, in the universe there are more atoms of hydrogen than there are of all the other elements put together. The Sun is made up largely of hydrogen—at least 70 per cent—and it is this hydrogen which acts as 'fuel' and makes the Sun shine.

Near the centre of the Sun, the temperature is very high. It reaches at least 14 million degrees Celsius, and may be rather more. The pressure also is tremendous. The result is that the atoms of hydrogen are running together to build up atoms of the next lightest element, which is known as helium (pronounced 'hee-lee-um'). It takes four atoms of hydrogen to make

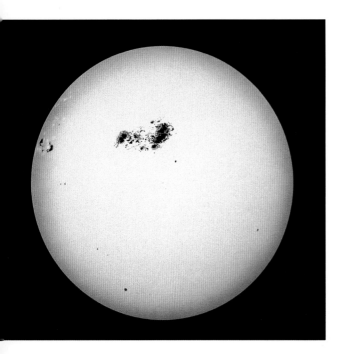

This artist's impression, drawn in 1989, shows a large sunspot group which lasted several weeks before disappearing. Sunspots are cooler than the surrounding bright surface, which is why they look dark.

16

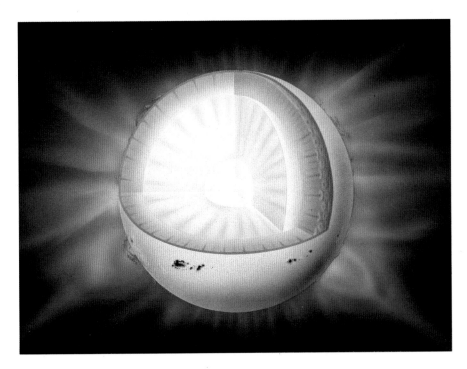

(*Left*) The Sun's 'power-house', where its energy is produced, is at the centre of the huge globe.

(*Below*) In this comparison of the size of the Sun with that of Betelgeux, the Sun is the small disk on the lower right. Because Betelgeux is so immense, only part of it can be shown.

up one atom of helium. Each time this happens, a little energy is set free and a little 'mass' (or weight, if you like) is lost. It is this energy which keeps the Sun shining, and the loss in mass is as much as 4 million tons every second. The Sun 'weighs' much less now than it did when you started reading this page. Luckily, there is no need to be alarmed; the Sun has so much mass that it will not change much for at least 5000 million years in the future.

The building-up of helium is done in a rather roundabout way, particularly as all the atoms near the Sun's centre are broken up, but the end result is quite definite. As time goes by, more and more of the Sun's hydrogen is turned into helium.

Not all the stars have the same temperature, as we can tell from their colours. For example, the Sun is yellow, while Betelgeux, the bright star in Orion, is red. This means that Betelgeux has a cooler surface; remember that when you switch on an electric fire you find that the bars first glow red and then change to orange, yellow and white as they become hotter. White stars, such as Rigel and Sirius, are therefore hotter than the Sun. From all this information, we can find out a great deal about the life-stories of the stars.

RELATIVE SIZE OF BETELGEUX AND THE SUN

Betelgeux

Sun

CHAPTER 7
GIANTS AND DWARFS OF THE SKY

More than 70 years ago, two astronomers, Ejnar Hertzsprung of Denmark and Henry Norris Russell of the United States, drew up what are still called Hertzsprung-Russell (or HR) diagrams. What they did was to plot the stars according to their surface temperatures, and their real luminosities, taking the Sun as being equal to 1. A typical HR diagram is shown here. The Sun has a surface temperature of 6000 degrees Celsius, so we can enter it on the diagram. Sirius has a slightly higher temperature, and is 26 times as powerful as the Sun; Betelgeux is much cooler, but is equal to 15,000 Suns, and so on. I have put in a number of famous stars which you can see on a clear night, as well as others which are too faint to be seen without a telescope.

You will notice one fact at once. There is a band in the diagram, from top left to bottom right, in which we find most of the stars. This is called the 'Main Sequence'. The Sun is an ordinary Main Sequence star.

Notice, too, that the cool red stars are either very much more luminous than the Sun, or else very much fainter. They are either giants or dwarfs; red stars of about the same power as the Sun are missing. If a red star is very powerful, it must be very large—for example, Betelgeux has a diameter of 200 million miles (320 million kilometres), big enough to swallow the whole path of the Earth round the Sun. If you could drive round it, moving at a steady 100 miles (160 kilometres) per hour, it would take you almost a thousand years to complete the journey and come back to your starting point. On the other hand, the feeble red dwarfs are much smaller than the Sun.

I have already mentioned nebulae, which are clouds of gas and dust in space. Not many are visible with the naked eye, but you can see one below Orion's Belt, and many more can be seen with the aid of a telescope. Most of the gas is made up of hydrogen. Nebulae are very large indeed, and stars are born inside them.

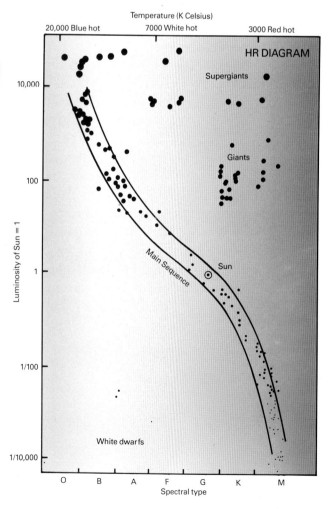

The letters at the bottom show the spectral types, O being the hottest stars and M the coolest. Luminosity is shown in the left-hand scale; the Sun is taken to be 1.

At first the star is very large and cool, but as it shrinks, because of the force of gravity, it heats up so much that it starts to shine. When the temperature at the star's centre has reached about 10 million degrees Celsius, hydrogen atoms run together to make up helium, and the star settles down at a point on the Main Sequence. This is what has already happened to the Sun.

There is so much hydrogen that the star goes on shining steadily for a long time—in the case of the Sun, about 10,000 million years—but the hydrogen will not last for ever. When it starts to run out, the Sun will have to use different 'fuels' to build up heavier and heavier kinds of atoms. The inside will shrink and become even hotter, while the outer layers will swell out and cool down. The Sun will leave the Main Sequence and turn into a red giant, sending out at least a hundred times as much energy as it does now. Next, the

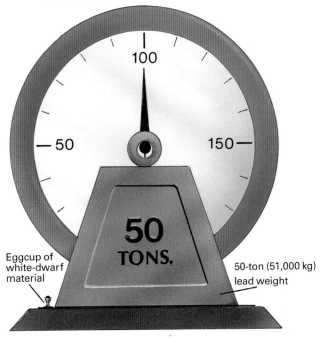

DENSITY OF WHITE-DWARF MATERIAL

A white dwarf is so dense that an eggcupful of material from one would weigh 50 tons.

The nebula which can be seen below Orion's Belt. Nebulae are immense clouds of dust and gas in which stars are born.

outer layers will be puffed off altogether, and for a while the Sun will be surrounded by a 'shell' of thin gas; this is what we call a 'planetary nebula', though it has nothing to do with a planet and is not really a nebula.

When all the Sun's 'fuel' is used up, there will be another change. All that is left of the Sun will shrink into a very small body known as a white dwarf, in which all the atoms are broken and packed tightly together, so that the material is very dense. If you could fill an eggcup with white-dwarf material, it would weigh 50 tons. The white dwarf will shine feebly for another long period, but at last it will turn into a cold, dead globe.

We know of many of these feeble, dying stars; the brilliant Sirius has a white-dwarf companion. The Sun must turn into a white dwarf when its energy has been used up, but at present it is a steady, well-behaved star.

CHAPTER 8
EXPLOSIONS IN THE SKY

So far, I have been talking about stars of the same kind as the Sun; but with stars of much greater or lesser mass (that is to say, 'heavier' or 'lighter' stars) things are different. A very low-mass star will never become hot enough to use hydrogen as a fuel; it will simply shine as a dim red dwarf and then fade away. But what about a star which begins by being much more massive than the Sun?

The star goes through its Main Sequence period in only a few millions of years before turning into a red giant, with a central temperature as high as 100 million degrees Celsius. Heavier and heavier atom-groups are built up, and at last the core is made chiefly of iron.

This means disaster. Iron cannot be used as a fuel in the same way as hydrogen or helium. There is no more energy, and gravity makes the star collapse. As the outer parts crash down upon the iron core, there is a tremendous explosion. In only a few seconds the star blows most of itself away into space in what is called a *supernova* outburst, and for a few days or weeks it may send out as much power as a hundred million Suns put together. At the end of the outburst, all that remains is a patch of expanding gas, together with a very small object, only

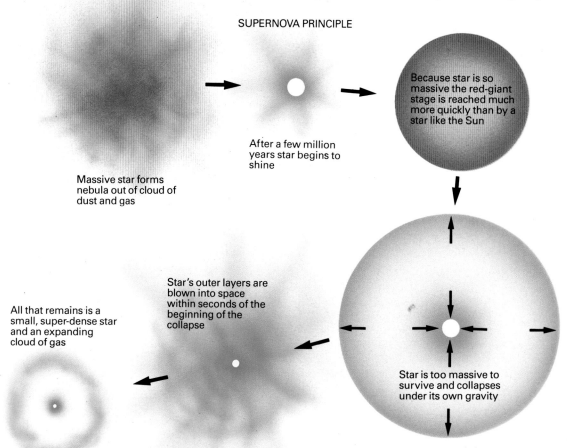

SUPERNOVA PRINCIPLE

Massive star forms nebula out of cloud of dust and gas

After a few million years star begins to shine

Because star is so massive the red-giant stage is reached much more quickly than by a star like the Sun

Star is too massive to survive and collapses under its own gravity

Star's outer layers are blown into space within seconds of the beginning of the collapse

All that remains is a small, super-dense star and an expanding cloud of gas

The Crab Nebula, an immense patch of expanding gas. Its light takes 6000 years to reach us.

a few miles or kilometres across, which is simply the old core of the dying star.

In the year 1054 a supernova was seen in the constellation of Taurus (the Bull), not far from Orion in the sky. For a few months it was bright enough to be seen with the naked eye in broad daylight. It then faded away, but with our telescopes we can still see its remains as a patch of gas—known as the Crab Nebula, because it was once said that the shape of the cloud is rather like a crab. Remember, though, that it is 6000 light-years away, so that the actual supernova outburst happened well before there were any astronomers on Earth to study it. In our Galaxy, only two supernovae have been seen since then, one in the year 1572 and the other in 1604, but we can trace the remains of other supernovae, mainly because they are still sending out radio waves.

In 1968, astronomers at Cambridge University in England were using a radio telescope when they suddenly found a strange source which seemed to be 'ticking' every few seconds. They did not know what it was, and for a time they even thought that it might be a signal sent to us from someone living on a planet moving round another star. (People even started talking about 'LGM' or Little Green Men.) But before long, it was found that the ticking signals came from the remains of a supernova.

If the parts of broken-up atoms are squashed together with enough force, they will make what are called *neutrons*. Because there is no waste of space, neutron material becomes very dense, so that a pin's head of it would weigh as much as the liner *Queen Elizabeth II*. The core of an old supernova is made up of neutrons. As we have seen, it is very small, but it is as massive as the Sun, and it is spinning round quickly—30 times every second in the case of the neutron star in the Crab. As it spins, it sends out beams of radio radiation, rather in the manner of the rotating beam of a lighthouse. Each time a beam sweeps across the Earth we pick up a 'pulse' of radio energy, and so these neutron stars are usually called *pulsars*. Since the first Cambridge discovery, many others have been found, so that there have been many supernovae in the Galaxy in the past.

Nobody will ever try to land on a pulsar. If you could do so, you would find that you would weigh millions of millions of tons, and it would not be very comfortable!

Of one thing at least we can be sure: the Sun will never explode as a supernova. It is not nearly 'heavy' enough, so that it will end its life-story as a white dwarf instead of turning into a pulsar.

CHAPTER 9
BLACK HOLES

A star such as the Sun will end up as a white dwarf; a more massive star will explode as a supernova. With a star which is 'heavier' still—say at least eight times as massive as the Sun—the story is different again.

As before, the star is born inside a nebula, and goes through its Main Sequence and giant stages. But when all the energy has been used up, the star collapses so suddenly and so violently that nothing can stop it. The star does not even have time to explode as a supernova. It becomes smaller and smaller, denser and denser; and as it does so, its *escape velocity* increases.

If you throw an object upwards, it will rise to a certain height and then fall down. Throw it harder, and it will rise higher before dropping back. If I could throw an object up from the Earth at a speed of 7 miles (11 kilometres) per second, or around 25,000 miles (40,000 kilometres) per hour, it would never come down at all; the Earth's gravity would not be strong enough to hold it, and the object would escape into space. This is why 7 miles per second is called the Earth's escape velocity (and also why spacecraft sent to the Moon or the planets have to work up to this speed).

If the Earth were more massive, it would pull more strongly, and the escape velocity would be greater. The Sun, which is over 300,000 times 'heavier' than the Earth, has an escape velocity of 383 miles (617 kilometres) per second. The greater the mass, and the smaller the body which is pulling, the higher will be the escape velocity.

Now let us go back to our collapsing star. When it becomes small enough, the escape velocity will become as great as 186,000 miles (300,000 kilometres) per second. This is the speed of light—and light is the fastest thing in the universe; nothing can possibly go any quicker. Therefore, not even light can escape from the collapsed star. It is pulling so strongly that it has surrounded itself with a region from which nothing can break free. It has become a *black hole*.

We cannot see a black hole, because it sends out no energy at all, but it still has gravity, and this can affect visible bodies close to it. A black hole is a cannibal. It gobbles up any material close to it, and once this material is

ESCAPE VELOCITY

Less than 5 mps (8 kmps) – object falls back to Earth

More than 7 mps (11 kmps) – object escapes Earth and never returns

5 mps (8 kmps) – object goes into orbit around Earth

Earth

If an object is fired away from the Earth at less than 5 miles (8 km) per second, it will fall back; if it is sent up at 5 miles per second, it will go into a path round the Earth: if it is sent up at over 7 miles (11 km) per second it will escape the Earth's pull.

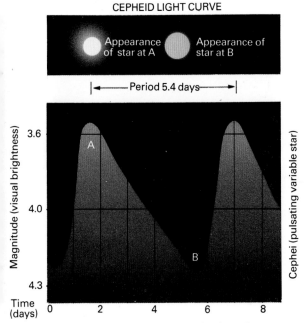

Light curve of Delta Cephei – the best-known Cepheid variable, with a period of 5.4 days. The magnitude changes from 3.4 to 4.2.

Cepheids (pronounced 'see-fee-ids'). It had already been found that the period of a Cepheid—that is to say, the time taken between one maximum brightness and the next—tells us how luminous the star really is. The longer the period, the more powerful the star. If we know how bright a Cepheid looks, and also know how luminous it really is, we can work out its distance.

The best comparison I can give is to picture a faint light seen from the seashore out across the water. If we know that it is really brilliant, it must be a long way away. If it is a dim light on a fishing-boat, it will be close to the shore.

Cepheid variables are found in globular clusters. Shapley looked at them, found their distances, and was then able to draw up a map of the Galaxy itself. We now know that the full diameter of the Galaxy is around 100,000 light-years.

Shapley also found that the globular clusters are not spread evenly all over the sky. Most of them lie in the southern part. This is because the Sun, with its planets, lies well away from the centre of the Galaxy, so that we are having a lop-sided view. The distance between the Sun and the centre of the Galaxy is about 30,000 light-years.

If you could look from 'above' or 'below', you would see that the Galaxy is spiral in shape, like a Catherine-wheel, with the Sun lying near the edge of one of the spiral arms. Not surprisingly, the Galaxy is spinning round. The Sun takes about 225 million years to make one full journey around the centre—a period sometimes called the *cosmic year*. One cosmic year ago, the most advanced life-forms on Earth were amphibians, the ancestors of our frogs and toads; even the giant dinosaurs had yet to appear. We may wonder what conditions on the Earth will be like one cosmic year from now!

A cosmic year is the time it takes the Sun to make one full journey round the centre of the Galaxy – 225 million years. Dinosaurs existed less than one cosmic year ago.

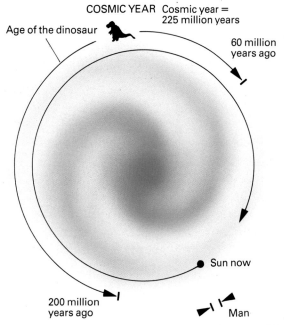

CHAPTER 11
OTHER GALAXIES

More than 200 years ago, a French astronomer named Charles Messier spent night after night studying the sky. He did not have a large telescope, but he was a very good observer. He was interested mainly in comets, which, as we have seen, are members of the Solar System. Messier discovered many comets, but he also came across star-clusters and nebulae which looked very like comets, and which wasted a great deal of his time. At last he decided to make a list of them, so that he could look them up straight away and make sure that he did not mistake them for comets. In 1781 he published a list of more than a hundred of them, and we still use the numbers he gave. For example, the great nebula in Orion is Messier 42, or M42 for short; the Crab Nebula in the Bull is Messier 1.

Star-clusters of all kinds were listed in the catalogue, as well as gaseous

The Whirlpool Galaxy in the constellation of the Hunting Dogs. Its light takes 37 million years to reach us.

The Trifid Nebula, in the constellation of the Archer, is a region where new stars are being born. Note the dark dust patches.

nebulae, but there were also objects which did not look like gas-patches. Messier 31, in the constellation of Andromeda, looked much more as though it were made up of stars. Messier himself did not care one way or the other, but later astronomers did their best to find out what these 'starry nebulae' really were.

In 1845 the Earl of Rosse, in Ireland, built what was then much the largest telescope in the world. It collected its light by using a mirror 72 inches (183 centimetres) across. Lord Rosse looked at the starry nebulae, and found that some of them were spiral. One of them, Messier 51, is face-on to us, and has been nicknamed the Whirlpool. The brightest of all the starry nebulae,

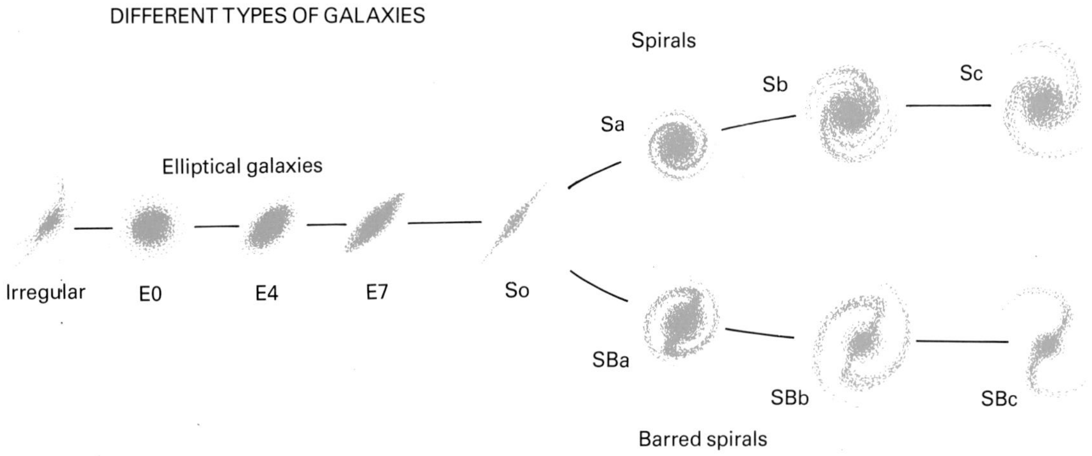

DIFFERENT TYPES OF GALAXIES

Types of galaxies – irregular, elliptical, spiral and barred spiral.

Messier 31, is also spiral, though it lies almost edge-on to us and we cannot see the spiral form properly.

Were the spirals members of our Galaxy, or were they much further away, in which case they would be galaxies in their own right? Astronomers did not know, and it was difficult to find out, because the spirals were so far away.

Then, in 1923, the American astronomer Edwin Hubble made careful studies with the help of the great reflecting telescope at Mount Wilson, in California. This has a mirror 100 inches (254 centimetres) across, and was much more powerful than Lord Rosse's telescope. Also, Hubble could make use of photography.

In some of the spirals, including Messier 31, he found Cepheid variables. By measuring their changes in light, Hubble was able to find out how far away they were, and at once he saw that they lay well beyond the edge of our Galaxy. The spirals were indeed separate galaxies. We now know that Messier 31 is over 2 million light-years away, and is larger than our own Galaxy, containing more than our own 100,000 million stars.

Many millions of galaxies are now known, though not many are bright enough to be seen with a small telescope. By no means all of them are spiral. Some are globes, rather like globular clusters even though they are much larger. Some are elliptical. (An ellipse is rather like a stretched-out circle.) Other galaxies are irregular and have no shape at all. Hubble divided them into various classes, as shown here; note the strange 'barred spirals', in which the spiral arms seem to come from the opposite ends of a 'bar' through the centre of the system.

Though some galaxies, including Messier 31, are much larger than ours, others are smaller. Messier 31 has two fainter companions, both elliptical. One of them is in Messier's list, as M32; the other is not, but was given in the New General Catalogue of clusters and nebulae drawn up in 1887. We call it NGC 205.

Hubble's work showed that the universe is far larger than most people have believed. This was probably the most important discovery made by astronomers since the invention of the telescope.

CHAPTER 12
DIFFERENT KINDS OF GALAXIES

Though we know so much about the life-stories of the stars, we have not found out nearly so much about the life-stories of the galaxies. It was at first thought that a spiral galaxy might turn into an elliptical system, or an elliptical into a spiral, but this does not seem to be true.

We do know, however, that the spirals are places where stars are still being born, because the brightest stars in them are hot and blue or white, so that they have not grown old enough to turn into red giants. There is also a large amount of dust and gas spread between the stars of a spiral, particularly in the arms, and there are also many gaseous nebulae. Elliptical galaxies are different. Their brightest stars are old and red, and all the star-forming dust and gas has been used up, so that they have gone further in their life-stories than the spirals.

Galaxies often collect together in clusters or groups. (Remember, a cluster of galaxies is quite different from a globular star-cluster.) Our Galaxy is a member of such a group; we call it the 'Local Group', because it is close to us on the astronomer's scale. The largest member of the group is Messier 31, the Andromeda Spiral. Next in size comes our own Galaxy. There is also a smaller spiral in the constellation of Triangulum (the Triangle), which is 2,900,000 light-years away, and there are more than 20 smaller systems, some of which are so faint and poor that they hardly look like galaxies at all.

The two nearest fairly large galaxies are known as the Clouds of Magellan, because they were seen by the famous Portuguese explorer Ferdinand Magellan during his voyage round the world in 1520. In fact they must have

LOCAL GROUP OF GALAXIES

(*Left*) Our Milky Way Galaxy is a member of the Local Group, with its two companions, the LMC and SMC (the Large and Small Clouds of Magellan). (The circles are at intervals of 1 million light years).

(*Right*) The Small Cloud of Magellan is about 190,000 light-years away. It is too far south in the sky to be seen from Britain, but it is easily seen with the naked eye from countries such as Australia.

This photograph of the Virgo Cluster, taken with the Palomar telescope, shows many galaxies. This cluster, over 40 million light-years away, contains many hundreds of systems.

been seen much earlier, because they are bright naked-eye objects, and look rather like broken-off parts of the Milky Way. Unfortunately they are so far south in the sky that they can never be seen from any part of Europe. Both are between 170,000 and 190,000 light-years away, and both are much smaller than a Galaxy.

The Clouds are important because they contain bodies of all kinds—giant and dwarf stars, loose clusters of stars, globular clusters and nebulae, as well as gas and dust. In 1987 a supernova blazed out in the Large Cloud, and became almost as bright as the Pole Star for a few weeks. I well remember seeing it; it was strange to think that I was looking at an explosion which really happened 170,000 years ago.

Further away, we see other groups of galaxies. One cluster, situated in the

constellation of Virgo (the Virgin) contains hundreds of systems, so that it is much larger than our Local Group, but it is at least 50 million light-years away, so that its members are too faint to be seen without a telescope.

The galaxies in the Local Group are close enough for us to see separate stars in them, and indeed we can still see stars in galaxies as much as 100 million light-years away. Galaxies further from us than this are much more difficult to study, and with systems thousands of millions of light-years away we can see no stars at all; the galaxies look like nothing more than smudges of light, though we can often tell whether or not they are spiral.

Edwin Hubble first measured the distances of galaxies by watching the variable stars in time. But if we cannot see separate stars, we must find another method. Supernovae often appear in galaxies, and because we have a good idea of how luminous these supernovae are, we can use them to estimate distances. When even the supernovae are lost in the distance, we have to turn to the spectroscope—and this brings us back to the Doppler Effect and the red shift.

CHAPTER 13
THE EXPANDING UNIVERSE

I have already said something about the Doppler Effect. If a star is coming towards us, its light is a little 'too blue', and all the lines in its spectrum are moved over to the blue or short-wave end of the rainbow band; if the body is moving away, the spectral lines are shifted towards the red. The spectrum of a galaxy is made up of the spectra of millions of stars jumbled together, but the main lines can be seen, so that we can find out how the galaxies are moving.

This was first done by American astronomers in 1912. To their surprise, they found that apart from a few nearby systems, those which we now know to belong to the Local Group, all the galaxies showed red shifts, so that all of them were moving away from us.

At the time nobody could understand this, and it was not even known that the galaxies are separate systems rather than parts of our own Galaxy; this was not discovered until Hubble's work more than ten years later. But when it was shown that the galaxies really are separate, things became even more strange. It looked as though our Galaxy were in the centre of the universe, with all the other galaxies moving outwards from it in all directions.

This is not true. We are not in a special position in the universe. All the groups of galaxies are moving away from all the other groups; the whole universe is spreading out—*expanding*.

Picture a cake in which there are numbers of currants. If the cake swells, each currant will move away from each other currant, as shown here. If the cake represents 'space' with galaxies spread through it, we can see the same sort of effect; as space expands, so the galaxies become further and further apart.

Notice, too, that there is no absolute centre to the universe, and that all the movements are 'relative', as was shown by the great German scientist Albert

(*Below left*) The spectrum of a star.
(*Below right*) The spectrum of a galaxy.

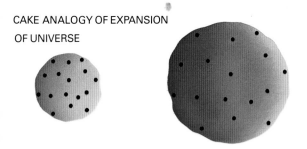

CAKE ANALOGY OF EXPANSION OF UNIVERSE

If a cake is expanded, the currants will move away from each other. If space expands, the galaxies move apart.

Einstein. If we lived in, say, a galaxy which is moving away from our own Galaxy at 1000 kilometres per second, it would seem to be *our* galaxy which was moving outwards. No matter where we happen to be in the universe, the general picture will be the same.

Hubble also found that there is a connection between a galaxy's speed, and its distance from us. The further away it is, the faster it is going. This in turn gives us a way of finding out the distances. We measure the red shift in the lines of the spectrum; once we know this, we can tell how fast the galaxy is moving away, and from Hubble's Law we can then work out its distance. In this way, we have been able to show that some galaxies are well over 10 thousand million light-years away, and are moving out at more than 90 per cent of the speed of light.

The force of gravity tends to pull everything together; why, then, are the galaxies spreading out instead of rushing together? The reason is that space itself is expanding (in the same way as the cake in our picture), carrying all the galaxies with it.

We have to admit that all our ideas depend upon one thing: the Doppler Effect. If the red shifts in the spectra of the galaxies are due to some other cause, we will have to start thinking all over again. There are some astronomers who believe this. Dr Halton Arp, in Germany, has found that there are galaxies which are close together, and are joined by 'bridges' of dust and gas, but which show quite different red shifts, in which case all our distance-measurements may be wrong. Not many astronomers agree with him, but we cannot be sure, and there is always the chance that we will have to change many of our present-day ideas.

Over the same period of time, a more distant galaxy will seem to move away from us more quickly than a galaxy which is closer to us.

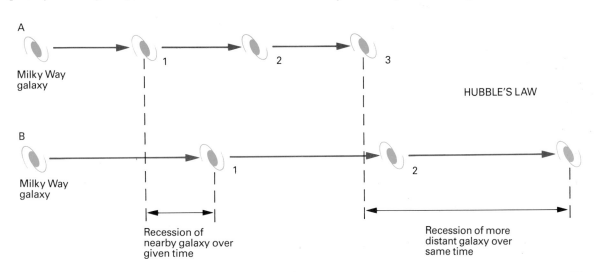

HUBBLE'S LAW

Recession of nearby galaxy over given time

Recession of more distant galaxy over same time

CHAPTER 14
THE STORY OF QUASARS

I have already talked about radio telescopes, which are not true telescopes at all, but are more like large aerials. The Lovell Telescope in Britain has a huge 'dish' 250 feet (75 metres) in diameter; it collects the radio waves and brings them to focus, just as an ordinary telescope collects and focuses light-waves, but instead of giving a picture it produces a trace on a graph. This shows how powerful the radio waves are, and where they come from.

It was soon found that some galaxies are very strong radio sources, perhaps because there have been violent explosions inside them. It was also found that radio waves came from positions in the sky where there did not seem to be any galaxies. Some of these sources

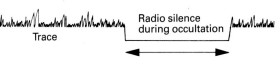

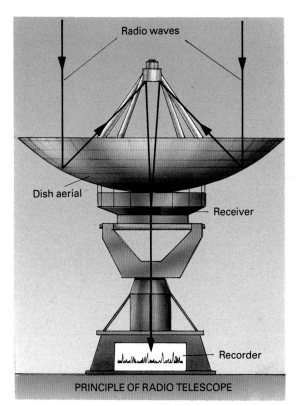

The principle of occultation. When the Moon passes in front of a radio source the radio signals are cut off. As the position of the Moon at this moment is known, the position of the radio source can be found.

agreed with the positions of what looked like faint blue stars—but why should stars of this kind be radio sources?

Then came a stroke of luck. As the Moon moves across the sky, it sometimes passes in front of a star and hides it. Since the position of the Moon at that time is known very accurately, we can also tell the position of the star, because we can measure the exact moment when its light is cut off. In 1963 it was found that the Moon would pass in front of a radio source known as 3C–273, because it was the 273rd object in a list of radio sources drawn up at Cambridge University. Astronomers in Australia were able to time the moment when the radio waves were cut off, and to show without any doubt that they really did come from the faint blue star.

A radio telescope of the 'dish' type. The radio waves from space are collected by the dish aerial, brought to focus and sent to a receiver.

The quasar 3C–273, believed to be the centre of a very active galaxy. It is very remote, and super-luminous.

The results were sent to the Palomar Observatory in California, where the largest telescope is a huge reflector with a mirror 200 inches (508 centimetres) across. At Palomar, Dr Maarten Schmidt used the telescope to study the spectrum of the 'star'. When he did so, he found to his surprise that the spectrum was quite unlike that of an ordinary star. There were lines in it, but he could not decide what caused them.

After some time, he found the answer. The lines were due to ordinary hydrogen, but they showed a very large red shift. This meant that 3C–273 was moving away very quickly, and must therefore be several thousands of millions of light-years away. It is not a star at all, but something much more important. Nowadays we call it a *quasar* (pronounced kway-zar).

Since then, hundreds of quasars have been tracked down, though by no means all of them send out powerful radio waves. For many years nobody could understand just what they were. They were much smaller than galaxies, and yet they were much more powerful. One quasar may send out as much energy as a hundred galaxies put together—and remember, a galaxy such as ours contains at least 100,000 million stars.

What could cause this amazing amount of energy? It was suggested that there might be supernova explosions going on all the time, but even this would not give enough power. Could the explanation be that a quasar contained a very large black hole, which was sucking in whole stars and destroying them? Or could there be some other cause, about which we knew nothing at all?

Today it seems almost certain that a quasar is the centre of a very active galaxy, but we have still not found out why it is so powerful. To make things even more puzzling, quasars change quite quickly in brightness, so that they cannot be very large. Though they have been known now for more than a quarter of a century, they remain almost as mysterious as they were when they were first discovered.

CHAPTER 15
THE EDGE OF THE UNIVERSE

One of the questions which I have been asked time and time again is: 'How big is the universe?' I always have to admit that I do not know—and neither does anybody else.

At the moment, the most distant quasars which we can measure are at least 13,000 million light-years away, moving outward at around 93 per cent of the speed of light. There are also some galaxies which are almost as far away. At least, then, we have a minimum size for the universe.

Edwin Hubble found a direct link between a galaxy's distance and its speed. If this so-called 'Hubble's Law' holds good, then we will finally come to a distance at which a quasar or a galaxy will be moving away from us at the full speed of light. In this case we

The Hubble Space Telescope, launched by the Space Shuttle in April 1990. It has a 94-inch mirror, and is much more powerful than any telescope on the Earth's surface.

will be unable to see it, and we will have come to the edge of the universe which is within our range. So far as we can tell, this limit is likely to be between 15,000 million and 20,000 million light-years.

This is all very well—but, remember, everything is *relative*. If we lived on a galaxy which is 13,000 million light-years from us, it would be our own Galaxy which would seem to be racing away at 93 per cent of the speed of light. It would be *our* Galaxy which would be near the edge of the observable universe!

In fact, all we can do is to give a limit to the size of the universe which we can ever hope to see. This is not the same thing as giving a limit to the size of the universe itself.

The trouble is that up to now we have not been able to 'see' as far as we would like. These very distant systems are so faint that they are visible only with our largest telescopes, and if they were much fainter we would not be able to see them at all. We cannot be sure that Hubble's Law will hold good at still greater distances.

What we need, of course, is a more powerful telescope. The telescope which was put into a path round the Earth in April 1990 may help, and may even be able to tell us whether or not there is a definite limit beyond which we can never see. It is surely right that this telescope has been named the HST or Hubble Space Telescope, in honour of Edwin Hubble.

There is another problem, too. We have been talking about the distances of the galaxies and quasars in space, but what about space itself? Is it really limited in size, or does it go on and on for ever?

If it has a definite size, then we want to know what lies outside it. If we answer: 'Nothing', we are no better off, because 'nothing' is simply space. But if space has no limit at all, we have to picture something which has no end, and this we are unable to do; our brains are not good enough.

One suggestion is that space may be 'finite, but unbounded'—that is to say, it has no edge but is still limited in size. The only way in which I can give any idea of what is meant is to ask you to

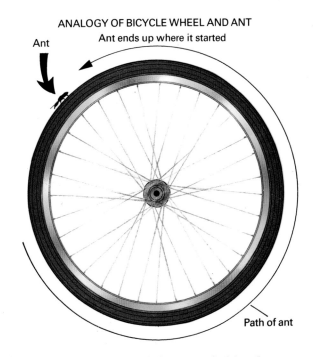

An ant crawling round the tyre of a bicycle wheel will never come to a boundary.

picture an ant crawling round the tire of a bicycle-wheel. If it crawls for long enough, it will come back to the place where it started, but it will have not found any 'edge' to the wheel.

Of course, this is not a proper comparison, because our crawling ant will have no idea of 'up' or 'down'. One thing is certain: the universe is a very large place indeed!

CHAPTER 16
THE BIG BANG

Our next problem is to find out the age of the universe. Either it began at one special moment, or else it has always existed, so that it was never 'born' in the usual meaning of the word.

At least we have some facts to guide us. The age of the Earth is known to be about 4600 million years, as we can tell from studying the make-up of the rocks. Since the Earth was formed from a cloud of dust and gas surrounding the Sun, we are sure that the Sun is older than the Earth, and there are many stars which are older than the Sun. The age of the Galaxy must be at least 10,000 million years.

We can do better than this, because we can see galaxies and quasars which are at least 13,000 million light-years away. The light we now see from them started on its journey 13,000 million years ago, so that the universe itself must be at least as old as that.

What we do not know, of course, is how all the matter in the universe was formed in the first place. The usual idea is that it came into existence at one moment, between 15,000 and 20,000 million years ago, in what is called the 'Big Bang'. Once it had appeared, the universe began to spread out. At first it was made up of hydrogen, which, as we have seen, is the lightest of all the elements and is still more common than everything else put together. As time went by, other types of atoms were built up from the hydrogen. Galaxies were formed; then came stars, some of which, such as the Sun, produced families of planets. Once we have a starting-point, we can work out the whole sequence of events, beginning with the Big Bang and ending up with you and me.

But where did the Big Bang happen?

(*Left*) Formation of the Earth by matter collecting together from a cloud of dust and gas around the young Sun.

(*Right*) The time-scale of the universe. On this scale the length of time for which Man has existed would be represented by 1/50 mm.

And what caused it? The answer to the first of these questions may sound strange. We have to say that space, time and matter all came into existence at the same moment. We cannot say 'where' the Big Bang happened. Since there was nothing else—not even space—the Big Bang happened 'everywhere'. This can at least explain why the present universe has no definite centre.

The second question is even harder to answer. We can work backwards to the tiniest fraction of a second after the Big Bang, but this is no help in explaining just how space, time and matter—indeed, the whole universe—could suddenly appear out of nothingness. It is best to admit that we simply do not know.

In 1948 a group of British astronomers, including Sir Fred Hoyle, put forward a different idea. They suggested that the universe has always existed, so that there was no Big Bang, and we can look back into the past for as far as we like. When a galaxy moves so far away that its light can no longer reach us, we will lose sight of it; but as old galaxies disappear, new galaxies are born from material which simply appears in the form of hydrogen atoms. If we could come back and look at the universe in, say, a million million years' time, we would see as many galaxies as

A remote quasar, indicated by the arrow. Its light takes over 10,000 million years to reach us.

we do now—but they would not be the same galaxies.

There is a way of checking whether this is true or not. When we look at galaxies which are thousands of millions of light-years away, we are seeing them as they used to be thousands of millions of years ago; we are looking back into the past. If the universe were in a 'steady state', as the British astronomers believed, then the very distant galaxies would be spread around in the same way as those which are closer to us—but this is not the case, and the whole picture is wrong. By now the steady-state idea has been given up. We are back to the Big Bang.

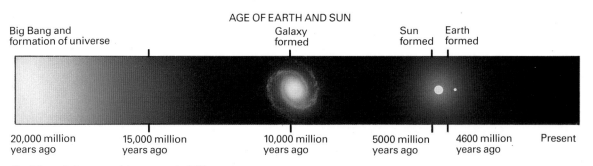

AGE OF EARTH AND SUN

Big Bang and formation of universe		Galaxy formed		Sun formed	Earth formed	
20,000 million years ago	15,000 million years ago	10,000 million years ago		5000 million years ago	4600 million years ago	Present

On this scale human existence equals 0.02 mm

CHAPTER 17
THE IN-AND-OUT UNIVERSE

At the present time the universe is expanding. All the groups of galaxies are racing away from each other. We must next try to find out whether this will go on for ever, or whether the galaxies will stop, turn round and come together again in what we may well call a 'Big Crunch'.

Everything depends upon the average density of material in the

A cluster of galaxies in the constellation of Coma Berenices (Berenice's Hair), photographed with the Palomar telescope.

universe—that is to say, how massive it is. If there is enough material, then the galaxies cannot fly apart until they lose touch with each other, because the force of gravity will pull them back, and we will have what we may call a 'closed' universe. If not, then there will be an 'open' universe, with expansion going on for ever.

We can see stars and galaxies, together with a tremendous quantity of gas and dust; but when we add all this up, we find that there is not nearly enough matter to pull the galaxies back. Unless there is more matter than we can see, the universe will be open rather than closed.

There does seem to be a large amount of this invisible 'missing mass'. We can tell this from the way in which the galaxies move in their groups. In some cases the groups ought to be breaking up, but they are not doing so. Something is holding them together, and this must be the missing mass—but what form is it in?

One idea is that the missing mass may be locked up inside black holes. Remember that although a black hole cannot be seen, because it is sending out no energy, its gravitational pull is as strong as if it were visible. If there are enough black holes, we can explain why the clusters of galaxies keep together.

Another suggestion is that there may be more faint red stars than we expect. Because they are so feeble, these red dwarfs can be seen only when they are fairly close to us. If, say, there were millions upon millions of extra red dwarfs in the Andromeda Spiral, we would have no hope of finding them, and if their numbers were great enough they might account for the missing mass.

Also, we have recently discovered large, massive galaxies which are so spread out, and so dim, that they have not been seen before. They may be as common as the bright galaxies, and this would make the amount of matter in the universe much greater than we have always thought. Finally, the missing mass may be in the form of matter which we cannot even begin to understand.

As yet we cannot decide one way or the other, and we must wait to see what

An artist's impression of a red dwarf, a low-mass star that never becomes very hot.

we can find out in the future. If the missing mass is sufficient, then we will have an in-and-out universe—big bang, expansion, big crunch, another big bang, and so on.

Yet even this does not help us to explain the birth of the universe. The atoms making up everything we can see must have come from somewhere or other, but we do not know how.

Let me try to give you a comparison. Go into a forest of oak trees; you will see saplings, young trees, and old trees many feet high. If you look at them, you will be able to work out how a tree grows; but if we did not know that it came from an acorn, we would be no better off—and in any case, what made the acorn?

At last we have some proof that there must have been a Big Bang. When it happened, the temperature must have been very high—many millions of degrees Celsius. As the universe spread out, with the expansion of space, it cooled down, but today we can still pick up weak radio waves, coming in from all directions, which are believed to be due to the after-effects of the Big Bang which happened so long ago.

CHAPTER 18
LIFE IN THE UNIVERSE

Now that we have taken a look at the universe, it is time to ask another question: 'Are we alone? Is there life anywhere except on the Earth?'

We can at once rule out almost all the members of the Solar System. None of them seems suitable for life, apart from our own world. The Moon and Mercury have no air; the atmosphere of Mars is too thin and the climate is too cold; Venus has poisonous clouds, and the giant planets do not even have solid surfaces on which you could stand. So we must look beyond the Sun's family if we are to have any hope of finding 'other men'.

Of course, I am talking about our own kind of life. Science-fiction writers are fond of inventing quite different forms, often termed BEMs or 'Bug-Eyed Monsters', which could live quite happily upon an airless world or inside a gas-giant. But all we know about science tells us that these alien life-forms cannot exist, and it is more sensible to keep to the facts as we know them.

If we are to find life elsewhere, we must look for another Earth-type planet. Unfortunately, no telescope yet made is powerful enough to show planets moving round other stars, but there is another way of showing that

IRAS, the Infra-Red Astronomical Satellite. Sent up in 1983, IRAS gathered information for nearly a year.

they probably exist. If we have a star which is much less massive than the Sun, moving round which is a planet as massive as Jupiter (the largest planet in our Solar System), the planet will pull upon the star and make it 'wobble' very slightly as it moves along in space. The wobbles will be very slight, but with our present instruments we could measure them and show that a planet exists. This has been done with several nearby stars.

We also have the results from IRAS, the Infra-Red Astronomical Telescope. This was a special telescope carried up in a man-made satellite which was launched in 1983 and was put into a path round the Earth, above the top of the air. As we have seen, infra-red radiation is sent out by a cool body which is not yet hot enough to shine in visible light. IRAS found that some stars are sending out so much infra-red that they must be surrounded by clouds of cool material which could make up planets.

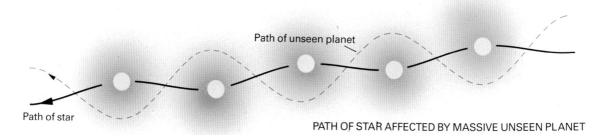

PATH OF STAR AFFECTED BY MASSIVE UNSEEN PLANET

Even if we have a world like the Earth, moving round a star like the Sun, we still cannot be sure that life will appear there. We are not even sure how life on our own Earth started, though it seems to have been in the warm seas which covered much of the world over 4000 million years ago. My own view is that life is likely to appear on any planet where conditions are right for it, but I may well be wrong.

Because we cannot send out rockets to planets of other stars, it seems very hard to think of any way of getting in touch with 'other beings'. The only chance seems to be by using radio, since radio waves move at the same speed as light (186,000 miles, or 300,000 kilometres, per second). We could send out a coded message which would be recognized as being different from a natural signal, and hope for a reply; we could also use our own radio telescopes to listen out. This has already been done, but without success.

Of course, it would be a slow business. The two nearest stars which are at all like the Sun, and might be expected to have families of planets, are 11 light-years away, so that any radio message would take 11 years to reach them and another 11 years to get back. Send out a signal in 1990, and you could not hope for a reply before the year 2012.

At the moment, this is about as far as we can go. Yet it is surely wrong to think that among all those stars in all those galaxies, only the Earth can support life.

(Right) An artist's impression of the Beta Lyrae system. Beta Lyrae is made up of two stars close together. This is the view that might be seen from a planet in the system.

(Left) A 'wobbling' star. If a low-mass star is being pulled by a massive planet, it will seem to 'wobble' very slowly and very slightly as it moves against the background of more distant stars. This wobbling will show how massive the unseen planet is.

CHAPTER 19
TRAVEL THROUGH THE UNIVERSE

We have sent men to the Moon and rockets out to the planets, but the stars are so far away that we cannot hope to send rockets there. The journey would take many thousands of years, and we could not even be sure that the star at which we aim would have a planet moving round it. Interstellar travel (that is to say, travel between the stars) is far more difficult than sending spacecraft through the Solar System.

Modern rockets are powered by liquids. In the future we may be able to use better fuels; but even so, we can never travel as fast as light. There is a definite reason for this. Anyone moving near the speed of light will find that his 'time' slows down, and if he could work up to the full speed of light his 'time' would stop.

Consider two twin boys, whom we will call Peter and John. John decides to go on a journey to a distant star, traveling in a spaceship which moves at over 90 per cent of the speed of light. If he goes to Sirius, he will have a journey of $8\frac{1}{2}$ light-years, so that by the time he has been there and back he will have been away for something like 20 years. But on his return, he will find that Peter has become an old man. In his fast-moving spaceship, John's time-scale has been slowed down. If he were away for, say, 50 years by his own time-scale, Peter would have died before John could return home.

This may sound strange, but we can show that it is true. The Earth is being hit all the time by particles called *cosmic rays*, which come in from all directions in space. When these cosmic rays crash into the upper air, they are broken up and produce showers of other particles which we call *mu-mesons* (pronounced 'mew-mee-zons'). These do not last for more than a tiny part of a second, and they ought to disappear before they have time to come down to the Earth's surface; but in fact they can reach the ground, because they are travelling so fast that their time-scale, compared with ours, is slowed down.

Of course, these effects are noticed only when you are moving at almost the speed of light. If you go on an aircraft trip or travel in a fast train, you need not worry!

Since we can never hope to move as fast as light, there is no chance that we

Star-clouds in the Milky Way, which is particularly beautiful when seen against a dark sky. Every speck of light in this photograph is a sun.

An artist's impression of our Galaxy as seen from the Magellanic Clouds. From a distance of 170,000 light-years it would show up as a spiral system.

can reach the stars by any method we know about today. And if we start talking about the ways in which Dr Who, Lord Darth Vader and the crew of the starship *Enterprise* manage it, we are back to the story-tellers. Yet it may happen one day; after all, nobody living a thousand years ago would ever have believed that we could reach the Moon, or look at a television screen and see a cricket match being played in Australia.

If we could look at the universe from a distant world, what would we see? From the Moon, the Earth is large and bright in the black sky, as the Apollo spacemen have told us. From Mars, the Earth would look like a bright bluish star, and telescopes would show its lands, its seas, its clouds and its ice-caps. But from even the nearest star, the Earth would be too small and faint to be seen at all, and the Sun would look like nothing more than a fairly bright point.

Go out to the Andromeda Spiral, at a distance of over 2 million light-years away, and even the Sun will be too dim to be seen; our Galaxy will be a smudge in the sky, rather smaller and fainter than the Andromeda Spiral looks to us. From the Virgo cluster, at 50 million light-years away, the Galaxy will have become a tiny speck; and from a quasar, the Galaxy will be beyond the range of telescopes as powerful as those we have built.

At least we have found out how small we are. We have learned a great deal since men believed the Earth to be the most important body in the universe.

CHAPTER 20
THE END OF THE UNIVERSE

We know that the universe was born at least 15,000 million years ago, and is probably rather older than that. What we must now do is to try to decide whether it will go on for ever, or whether it will come to an end.

We are in no danger! The Sun is shining steadily, and will go on doing so for thousands of millions of years. The chances of the Earth being hit by a solid body big enough to break it up, or do it real damage, are so small that we can forget about them, and there is no star close enough to hurt us even if it exploded as a supernova. In talking about the possible end of the universe, we are looking so far ahead that we can do little more than guess.

If the universe began with a Big Bang, started to spread out, and does not contain enough matter to pull the galaxies back, it will be rather like a clock which has been wound up and is now running down. As the groups of galaxies keep on moving outwards, they will finally lose all touch with

each other. The stars in them will die, and all that will be left will be cold, dead bodies which have lost all their energy.

But if there is enough matter in the universe—and we have already talked about the 'missing mass' which we believe must exist—then things are different. The galaxies will not go on moving outwards for ever. They will start to draw back towards each other, and after a very long time they will come together again in a Big Crunch, after which they will start to spread out once more. The universe will be rather like the musical instrument known as a concertina—in and out, in and out. In this case, the universe will not come to an end at all, and there will be new Big Bangs every 80,000 million years or so.

The Earth itself cannot last for ever. After it has used up its hydrogen 'fuel', in about 5000 million years from now, the Sun will swell out to become a red giant star, and will send out so much heat that the Earth will be destroyed. If men still live here, they will have to save themselves in some way. They may have found out how to move the Earth to a safer part of the universe; they may be able to leave the Earth altogether and move to another world—we cannot tell. This sounds like science fiction, but it is worth remembering that it is only a few thousand years since men lived in caves, so that if we manage to avoid fighting each other there is no limit to what we may be able to do in the far future.

Meanwhile, we must not be gloomy. The Earth is our home, and it lies in a part of the universe which is right for us. From it we can look out into space and see all the wonders of the universe—the stars, the great gas-clouds in which new stars are being born, the lovely spiral galaxies, and the strange, all-powerful quasars which are so far away that we now see them as they used to be long before our world was born.

This artist's impression suggests what might be seen from a planet near the centre of our Galaxy.

In this book I have tried to give you some idea of what astronomers of today are finding out about the universe. I hope that I have interested you. If so, then read books which will tell you more; there is always something new to find out. I wish you all success.

TEST YOUR SKILL!

Try to answer these questions. You will find all the answers in the book, but do your best before looking them up on page 48.

1. Why could you not live on Mars unless you stayed inside a spaceship or in a space-suit?

2. Look at the Sun and you will see it as it used to be $8\frac{1}{2}$ minutes ago. Why is this?

3. Which has the longer wavelength: red light or green light?

4. If a light-source is moving away from you, will its light seem a little too red or a little too blue? Why?

5. Is the star Betelgeux hotter or cooler than the Sun? How can you tell?

6. What is a nebula?

7. What is the main 'fuel' of the Sun?

8. How much 'weight' does the Sun lose every two seconds?

9. What is the name of the cloud of gas which is all that is left of the supernova seen in the year 1054?

10. Why can no light escape from a black hole?
11. Why are Cepheid variable stars so useful to astronomers?
12. The Andromeda Spiral is known as M31. What does the M stand for?
13. How far are we from the centre of the Galaxy?
14. How can we tell that the universe is at least 13,000 million years old?
15. Who first showed that the galaxies are outside systems rather than being parts of our own Galaxy?
16. If the star Sirius is $8\frac{1}{2}$ light-years away, how long would a radio signal from the Earth take to get there?
17. How long does the Sun take to complete one journey round the centre of the Galaxy?
18. How many stars are there in our Galaxy?
19. Who was the astronomer who first measured the size of our Galaxy?
20. Where is the Lovell Telescope?

ANSWERS

1. Because there is not enough air, and the climate is too cold. **2.** Because the Sun's light takes $8\frac{1}{2}$ minutes to reach us. **3.** Red. **4.** Too red—because its wavelength is lengthened: fewer light-waves per second reach you than they would do if the light-source were standing still. **5.** Cooler. We can tell this because Betelgeux is red, while the Sun is yellow. **6.** A cloud of gas and dust in space, inside which new stars are being born. **7.** Hydrogen. **8.** 8 million tons. (It loses 4 million tons every second.) **9.** The Crab Nebula. **10.** Because the old, collapsed star inside the black hole is pulling so strongly that its escape velocity is as great as the speed of light (186,000 miles, or 300,000 kilometres, per second). **11.** Because the way in which they change in brightness shows how luminous they are; and once this is known, we can find out how far away they are. **12.** Messier—the name of Charles Messier, who drew up a famous list of clusters and nebulae. **13.** About 30,000 light-years. **14.** Because we can see galaxies and quasars which are some 13,000 million light-years away. **15.** Edwin Hubble. **16.** $8\frac{1}{2}$ years. **17.** 225 million years (one 'cosmic year'). **18.** About 100,000 million. **19.** Harlow Shapley. **20.** At Jodrell bank, in England.